Contents

Building Research Establishment Report

Durability tests for building stone

K D Ross, BSc and R N Butlin, PhD, CChem, MRSC

Building Research Establishment
Garston
Watford
WD2 7JR

Price lists for all available
BRE publications can be
obtained from:
Publications Sales
Building Research Establishment
Garston, Watford, WD2 7JR
Tel: Garston (0923) 664444

BR 141
ISBN 0 85125 368 7

Introduction

Over the past few years there has been a considerable increase in the number of enquiries received by the Building Research Establishment about the durability or testing of natural stone. During this time it has also become apparent that the reasons why a particular test should be carried out are not generally understood by those who specify stone. This is not entirely surprising since there are very few British Standards which deal with the selection of stone for building.

The purpose of this report is to explain what tests should be carried out for particular stones, and to give details of the tests so that they may be carried out to a standardised procedure. The different types of stone require different tests, so they are dealt with individually. Table 1 summarises the tests required for these different types. This report does not however deal with the measurement of the physical properties of stone. Where a British Standard does exist details of the test are given for completeness. The most recent version of a Standard should be used because British Standards are revised periodically and the specification for the test may change.

Table 1   Summary of tests needed for different building stones

Type of stone	End use	Crystallisation test	Saturation coefficient	Acid immersion	Porosity	Wet/dry	Water absorption
Limestone	General	●	★	—	★	—	—
Sandstone	General	★	★	★	★	—	—
Sandstone	Severe exposure	●	—	●	—	—	—
Slate	Roofing	—	—	★	—	●	●
Slate	Copings	—	—	●	—	●	—
Slate	Damp course	—	—	●	—	●	—

● These tests should always be carried out for the stone in question
★ These tests may be required for certain applications of the stone — see text for details

Note The test conditions may vary for different stones; details are given in the main text

Existing British Standards

The following British Standards cover the use of stone in building.

BS 435:1975 Specification for dressed natural stone kerbs, channels, quadrants and setts
This specifies the use of igneous rocks (which include granites) but gives no information on testing.

BS 680:Part 2:1971 Specification for roofing slates
This specifies three tests which apply to roofing slate. They are: acid immersion, wetting and drying, and water absorption.

BS 743:1970 Specification for materials for damp-proof courses
This includes natural slate as a material for damp-proof courses. It does not include details of any tests but states that the slate should be tested in accordance with BS 5642, and shall be regarded as type B slate for the purposes of testing.

BS 5390:1976 (1984) Code of practice for stone masonry
This standard (formerly CP 121.201:1951 and CP 121.202:1951) deals with the use of natural stone in

ashlar and rubble-faced walls. A limited amount of advice on testing is given in this standard, but not enough to allow a testing programme to be specified. No information is included on what tests should be carried out on the various types of stone, and no practical details are given for any tests mentioned in the standard.

BS 5534:Part 1:1978 (1985) Slating and tiling. Part 1. Design

This deals with the use of natural slate as a roofing material from the design point of view. No tests are listed, but reference is made to BS 680 for testing.

BS 5628:1978 (1985) Code of practice for use of masonry

This standard is in three parts and covers the use of various materials (including natural stone) in the construction of masonry structures. No information on the durability testing of stone units is included, although the effects of design and environmental conditions on the overall durability of the structure are discussed.

BS 5642 Sills and copings
Part 1:1978 Specification for window sills of precast concrete, cast stone, clayware, slate and natural stone
Part 2:1983 Specification for copings of precast concrete, cast stone, clayware, slate and natural stone
Part 1 of this standard is a revision of BS 4374:1968, and Part 2 is a revision of BS 3798:1964. The standard specifies a sulphuric acid immersion test and a wetting and drying test for slate. Two grades of slate are recognised — A and B — depending on the degree of air pollution in the intended place of use. A different concentration of acid is used in the acid immersion test to distinguish between the two grades.

No test is specified for natural stone — the acid immersion test which was specified for sandstones in BS 3798 is not included in Part 2 of this standard.

BS 8298:1989 Code of practice for the design and installation of natural stone cladding and lining

This is a revision of CP 298:1972 and deals with the use of all stones as cladding material; there is no practical advice on testing.

BRE tests

Owing to the lack of British Standard tests for limestones and sandstones, BRE has developed its own tests which have now become accepted by the industry as standard tests. These tests are listed below.

Crystallisation test
Crystallisation tests were first developed in the first half of the nineteenth century[1] as rapid simulated tests for frost susceptibility. Several versions are used throughout the world, but it is thought that the test specified later in the present report is the most appropriate one for the British climate. The main use

of the crystallisation test in Britain is for testing limestones, but it can be of limited use for sandstones.

Crystallisation tests consist of exposing samples to alternate periods of soaking in a solution of a salt (usually sodium sulphate) and drying in a humid oven. During the test, samples lose weight according to their durability; the higher the weight loss, the lower the durability. For limestones the weight loss at the end of the test is used to assign a durability class.

The test recommended by BRE is a comparative one in that stones of known durability are included as internal reference samples. This provides a check to ensure that the test is giving consistent results. The test can also be used to compare two stones of unknown quality: in this case it is less important to include internal reference samples, but if they are excluded the results may not give a reliable figure for the calculation of durability class.

Saturation coefficient and porosity
The concept of saturation coefficient was developed by Hirschwald[2] and, like the cystallisation test, was also an attempt to devise a rapid test for frost susceptibility.

Saturation coefficient is defined as the ratio of the volume of water absorbed by a sample completely immersed in cold water for 24 hours, to the total volume of pore space in the sample. Hirschwald observed empirically that stones with a saturation coefficient of more than 0.8 tended to be susceptible to frost, but as a general rule the higher the saturation coefficient, the less durable the stone tends to be.

The test is much quicker to carry out than the crystallisation test, but the results are not so reliable. Nevertheless, the results do give a reasonable indication of durability for limestones.

Acid immersion test
This test is only carried out on sandstones. Its purpose is to identify those stones which are liable to decay when exposed to atmospheres highly polluted with acidic gases (eg industrial regions). The acid used to test stone for general building purposes is 20% (w/w) sulphuric acid. If a particularly long life is required of the stone, then a stronger solution of the acid (say 40%) may be needed.

Testing procedures for different types of stone

Limestones

There is no British Standard test for limestones. BRE recommends the following three tests for the routine testing of limestones:

(a) crystallisation test (using a 14% solution of sodium sulphate),

(b) saturation coefficient, and

(c) porosity.

The saturation coefficient and porosity can be calculated from the same set of data, and so will be described together.

Crystallisation test

Procedure

At BRE three limestones are used as internal reference samples: Portland Whit Bed, Box Ground and Monks Part (of good, moderate and poor durability, respectively). Box Ground is no longer available, but it is not particularly important which stones are used; it is more important that they cover the durability spectrum, are available over a period of years, and are of consistent quality during that period.

The test used at BRE is carried out as follows:

1   Make up a stock solution of sodium sulphate by dissolving 1.4 kilograms of sodium sulphate decahydrate in 8.6 litres of water, or by dissolving any hydrate of sodium sulphate in water until the specific gravity of the solution is 1.055 at 20°C. Approximately 2 litres will be needed to complete the test for each sample. The temperature of the solution should be maintained at 20 ± 0.5°C throughout the test.

2   Using a suitable saw, cut a representative number of 4 cm cubes of all the stones to be tested. No fewer than four cubes of each stone (including the internal reference samples) should be tested, but unless the stone is particularly variable a maximum of six should be sufficient.

3   Remove any loose material by washing with water, and dry the samples at 103 ± 2°C to constant weight (this can usually be achieved overnight in a ventilated oven of approximately 0.12 m³ internal volume).

4   Remove the samples from the oven, place them in a desiccator, and allow them to cool to 20 ± 2°C. Weigh them to ±0.01 grams (W_0).

5   Label the samples and weigh them again (W_1). (The best method of labelling is to attach a heat-resistant plastic label with wire and write the sample number on the label with waterproof ink.)

6   Place each sample in a 250 ml container and cover with the fresh sodium sulphate solution to a depth of about 8 mm. Leave for 2 hours during which time the temperature of the sample and solution should be kept at 20 ± 0.5°C. (Note: changing the soaking temperature can markedly affect the results — see Price[3].) When the samples have been soaking for 1½ hours, place a shallow tray containing 300 ml of water in the oven. (It was found empirically in the past that drying the samples in an oven which was initially humid improved the resolution of the test.)

7   After a total of 2 hours of soaking, remove the samples from the solution and put them in the oven on wire racks. Dry the samples for 16 hours* at 103 ± 2°C.

8   Remove the samples from the oven and allow them to cool to 20 ± 2°C. Steps 6 to 8 constitute one cycle of the test.

9   Repeat steps 6 to 8 until 15 cycles have been completed.

10 Weigh the samples (W_f). Calculate the percentage weight loss from the expression:

% weight loss = $100 (W_f - W_1)/W_0$.

11 Calculate the mean percentage weight loss for each set of samples.

Interpretation of data

The weight loss in the crystallisation test is used to allocate a durability class to each of the stones tested. This is done by reference to the first two columns of Table 2. Once a durability class has been assigned, the suitability of the stone for a particular position in a building under a variety of environmental conditions can be found by reference to Table 2 and Figure 1.

---

* If the test is to be halted for any reason (eg at weekends), leave the samples in the oven at 103 ± 2°C.

Table 2 Durability classes for limestones, based on weight loss in the crystallisation test, and the suitability of the stones for use in particular exposure zones of a building under different environmental conditions

Limestone durability class	Crystallisation loss (%)	Inland				Exposed coastal			
		Low pollution		High pollution		Low pollution		High pollution	
		No Frost	Frost	No Frost	Frost	No Frost	Frost	No Frost	Frost
		Zones #	Zones	Zones	Zones	Zones	Zones	Zones	Zones
A	<1	1 – 4	1 – 4	1 – 4	1 – 4	1 – 4	1 – 4	1 – 4	1 – 4
B	1 to 5	2 – 4	2 – 4	2 – 4	2 – 4	2 – 4	2 – 4	2* – 4	2* – 4
C	>5 to 15	2 – 4	2 – 4	3 – 4	3 – 4	3* – 4	4	—	—
D	>15 to 35	3 – 4	4	3 – 4	4	—	—	—	—
E	>35	4	4	4*	—	—	—	—	—
F	Shatters early in test	4	4	—	—	—	—	—	—

\# The exposure zones are illustrated in Figure 1
* Probably limited to 50 years' life

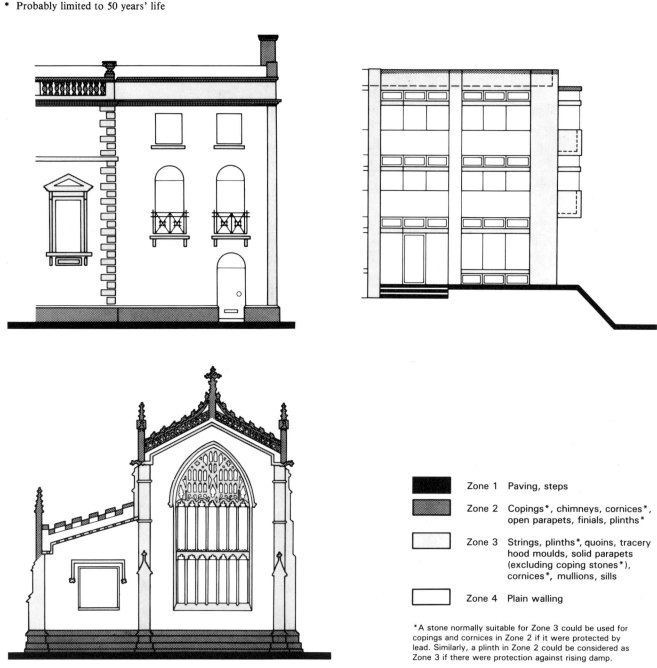

Zone 1   Paving, steps

Zone 2   Copings*, chimneys, cornices*, open parapets, finials, plinths*

Zone 3   Strings, plinths*, quoins, tracery hood moulds, solid parapets (excluding coping stones*), cornices*, mullions, sills

Zone 4   Plain walling

*A stone normally suitable for Zone 3 could be used for copings and cornices in Zone 2 if it were protected by lead. Similarly, a plinth in Zone 2 could be considered as Zone 3 if there were protection against rising damp.

Figure 1 Exposure zones of buildings

Saturation coefficient and porosity

Procedure

1 Dry a representative number of samples of the stone under test at 103 ± 2°C. The size and shape of the samples is not critical, but 4 or 5 cm cubes are normally used. Four samples of each type of stone are normally tested.

2 Allow the samples to cool to 20°C, and then place them in a vessel which is connected to a rotary vacuum pump and manometer, and has a tap funnel for the admission of water (see Figure 2).

3 Evacuate the vessel to a pressure of 3 mm of mercury or less, and maintain the vacuum for at least 2 hours.

4 Admit air-free water into the vessel until the samples are well covered, then admit air over the water to restore atmospheric pressure. Leave the samples under water for a least 16 hours.

5 Weigh the samples suspended in water (W_1), and again in air (W_2) after wiping off excess moisture with a damp cloth.

6 Dry the samples in an oven at 103 ± 2°C for 16 hours. Cool the samples and weigh them dry (W_0).

7 Immerse the dry samples in water at 15–20°C for 24 hours.

8 Remove the samples, wipe off excess moisture with a damp cloth, and weigh them (W_3).

9 Calculate the porosity P from the expression:

$$P = \frac{W_2 - W_0}{W_2 - W_1} \times 100 \quad (\%)$$

and the saturation coefficient S from:

$$S = \frac{W_3 - W_0}{W_2 - W_0}$$

Interpretation of data

Saturation coefficient and porosity should be used in conjunction with each other. Saturation coefficient is only of any use if the sample contains a significant amount of pore space. For example, a sample of Blue Clipsham stone was tested and found to be durability class A according to the crystallisation test, yet it had a saturation coefficient of 0.94. Normally, such a high value of S would indicate a stone of much lower durability. Similar results were obtained from a Purbeck marble: according to the crystallisation test results the stone was class A, but the saturation coefficient was 1.0 — the maximum possible. The reason for these apparently anomalous results is that the porosities of the two stones were quite low: 9.6% for the Clipsham, and 0.9% for the Purbeck.

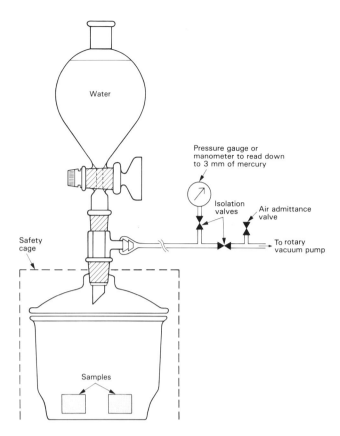

Figure 2 Diagram of apparatus for vacuum saturation of limestone samples with water

In practice therefore, saturation coefficient should only be used as a comparison between stones of similar porosity. Even then the results will not be totally reliable.

Sandstones

The majority of sandstones are resistant to frost, their general weathering behaviour being affected more by their chemical composition than by their structure. Thus, for sandstones which are to be exposed outdoors in a polluted environment, there is only one test that should be carried out routinely — that is the acid immersion test.

A 14% sodium sulphate crystallisation test can be of use if the sandstone is likely to be contaminated with soluble salts which could lead to decay. The test should only be used to compare the resistance of stones to salt attack, and never to allocate a durability class to sandstones as is done for limestones.

A saturated crystallisation test is sometimes used to assess performance in the severest locations when exceptionally long life is required (for example stone used for sea defence walls, quaysides, etc).

As far as sandstones are concerned, saturation coefficient and porosity are of no significance from the durability point of view; they may be of use in replacement work if the physical properties of the stone must be matched.

There are no British Standard tests for sandstones. The practical details for the crystallisation test, and the measurement of saturation coefficient and porosity are identical to those for limestones given above, except that the crystallisation test sometimes uses a saturated solution of sodium sulphate and sandstones are used for the internal reference samples.

Acid immersion test

Procedure
1 Prepare stocks of sulphuric acid of the correct strength (see Appendix).

2 Prepare six samples of each stone to be tested, each measuring $50 \times 50 \times 15$ mm. The samples should be dry.

3 Immerse each sample completely in 200 ml of the sulphuric acid solution in a covered vessel, and leave for 10 days.

4 Examine the samples for surface changes.

Interpretation of data
The acid immersion test is a pass/fail test. If the sample shows any sign of splitting or marked softening of the surface, then it is deemed to have failed. Loss of a few isolated grains does not constitute failure, however.

Slates

The testing of slates is covered in two British Standards: BS 5642 for slate coping units and sills, and BS 680 for roofing slates.

Slates for roofing

BS 680 specifies three tests for roofing slate: an acid immersion test, a wetting and drying test, and a water absorption test. A wetting and drying test and a water absorption test should be carried out on all slates, and an acid immersion test on those which are to be used in a polluted environment.

Sample preparation

Three specimens are required for each of the three tests, and each specimen should come from a different slate taken at random from the stock of slates.

1 Using a suitable tool (a small diamond saw is ideal) cut specimens of side 50 mm $\times$ 50 mm $\times$ the thickness of the slate, taking care to avoid cracking or splitting the slate.

2 Grind all edges to a smooth finish with fine abrasive and water.

3 Examine the specimens before testing; any specimens that have cracks or other defects shall not be used and another specimen shall be prepared.

Water absorption test

1 Dry the three specimens in an oven at 105°C for about 48 hours. Weigh the samples dry (W_0).

2 Immerse the samples in distilled water in a vessel fitted with a reflux condenser and boil gently and continuously for 48 hours.

3 Allow the samples to cool in air for 5 minutes and then place them in cold water and allow them to stand for 30 minutes in the room in which they are to be weighed.

4 Remove any surplus water from the samples by wiping with a damp cloth, and weigh them to the nearest 0.001 g (W_1).

5 Calculate the increase in weight of each specimen as a percentage of its dry weight using the expression

$$\% \text{ wt increase} = \frac{100 \times (W_1 - W_0)}{W_0} \%$$

The average of the three values is the water absorption of the slate. (Note that if the range of the three values is greater than 10% of the mean, the test may be repeated using another three samples, the average of the three new values being taken as the water absorption.)

A slate is deemed to have passed this test if the average water absorption of the three specimens is no greater than 0.3%.

Wetting and drying test

1 Immerse three specimens in distilled water at room temperature for 6 hours.

2 Dry the samples in a ventilated oven at 105 ± 5°C for 17 hours and allow them to cool to room temperature. Steps 1 and 2 constitute one cycle of the test.

3 Repeat steps 1 and 2 until the specimens show signs of splitting, flaking or delamination, or until 15 wetting and drying cycles have been carried out, whichever is less.

A slate is deemed to have passed this test if none of the specimens show signs of delamination or splitting along the edges when examined with a lens giving a magnification of about two diameters, nor of flaking of the surface.

Sulphuric acid immersion test

The practical details of this test are the same as those given earlier for sandstones, except that the acid strength is different (see Appendix) and there is no stipulation on the amount of acid that shall be used.

A slate is deemed to have passed this test if none of the specimens show delamination along the edges (when examined through a lens giving a magnification of about two diameters), or swelling, softening or flaking of the surface, or gaseous evolution during immersion.

Slates for sills and copings

BS 5642 specifies two tests for slate to be used for sills and copings: a wetting and drying test, and an acid immersion test. The Standard recognises two grades of slate — types A and B. Type A slate is suitable for areas of high pollution, whereas type B is only suitable for areas of slight to moderate pollution. Practical details of the acid immersion test are the same as those given earlier for sandstones, and details of the wetting and drying test are the same as those for roofing slates (see above) **except for the following changes.**

(a) Sample preparation
Two specimens of side 25 mm are cut from each of three slates selected at random, giving six specimens in all. The edges are ground to a smooth finish using fine abrasive and water. Any samples showing signs of splitting or any other defect are discarded and another sample prepared.

(b) Wetting and drying test
Six specimens are used when testing slate for use as copings.

(c) Sulphuric acid immersion test
Six specimens are used when testing slate for use as copings.

The strength of the acid is different (see Appendix) and not less than 1000 ml of acid are used separately for each specimen.

Slates for damp-proof courses

Slates for damp-proof courses are covered in BS 743. The tests carried out are exactly the same as those for sills and copings. The slate for damp-proof courses should be regarded as type B.

Granites and marbles

There are no standard tests for granites and marbles. BS 435:1975 'Specification for dressed natural stone kerbs, channels, quadrants and setts' specifies igneous rocks (which granites are) for these applications, but gives no practical advice on the selection of the rock. BS 8298 'Code of practice for the design and installation of natural stone cladding and lining' lists granites and marbles as suitable materials for cladding, again with no advice on selection except that the use of brecciated marbles is not recommended in certain cladding applications. If the suitability of these materials for a particular application is in doubt, then independent advice should be sought, or an inspection of the rock on an existing building carried out.

References

1 **de Thury H**, *et al*. On the method proposed by Mr Brard for the immediate detection of stones unable to resist the action of frost. *Annales de Chemie et de Physique,* 1828, **38** 160 – 192.

2 **Hirschwald J**. *Handbuch der Bautechnischen Gesteinsprüfung.* Berlin, Gebrüder Borntraeger, 1912.

3 **Price C A**. The use of the sodium sulphate crystallisation test for determining the weathering resistance of untreated stone. *Proceedings of the RILEM/UNESCO Symposium on Deterioration and Protection of Stone Monuments, Paris, June 1978.*

Appendix

Preparation of sulphuric acid for acid immersion tests

The dilute acid is prepared in a heat-resistant glass container by slowly adding the desired amount of concentrated (98%) sulphuric acid to the relevant amount of distilled water (see Table A1), stirring constantly. Great care should be taken in preparing the solutions because a lot of heat is generated during mixing. **On no account should water be added to concentrated sulphuric acid because under these conditions the heat released can cause the dilution process to become quite violent.** Protective clothing for the face, eyes and hands should also be worn (for more details see *Substances hazardous to health,* a loose-leaf reference book updated quarterly and published by Croner Publications Ltd of New Malden).

Table A1 Strength of acid for acid immersion tests

Type of stone	Test or Standard	Volume of 98% sulphuric acid	Volume of water
Sandstone	BRE 20%	300 ml	2155 ml
Sandstone	BRE 40%	300 ml	1015 ml
Roofing slate	BS 680	300 ml	2100 ml
Slate type A	BS 5642	20 ml	2370 ml
Slate type B	BS 5642	1 vol of acid prepared for type A slate in 4 vols of water	

8

Printed in the UK for HMSO. Dd.8157473, 4/89, C5, 38938.